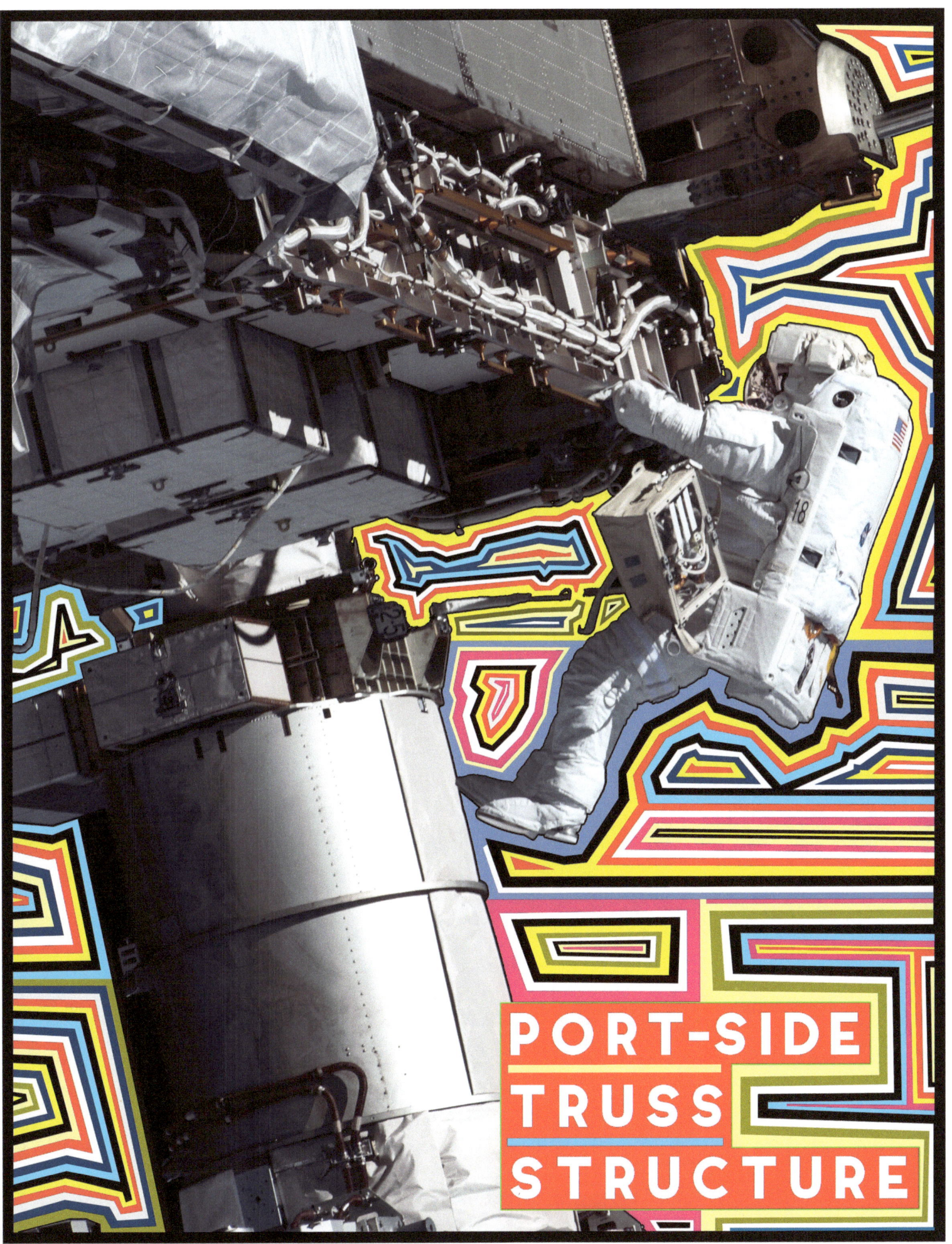

THE SOYUZ MS-17 CREW SHIP IS PICTURED DOCKED TO THE RASSVET MODULE

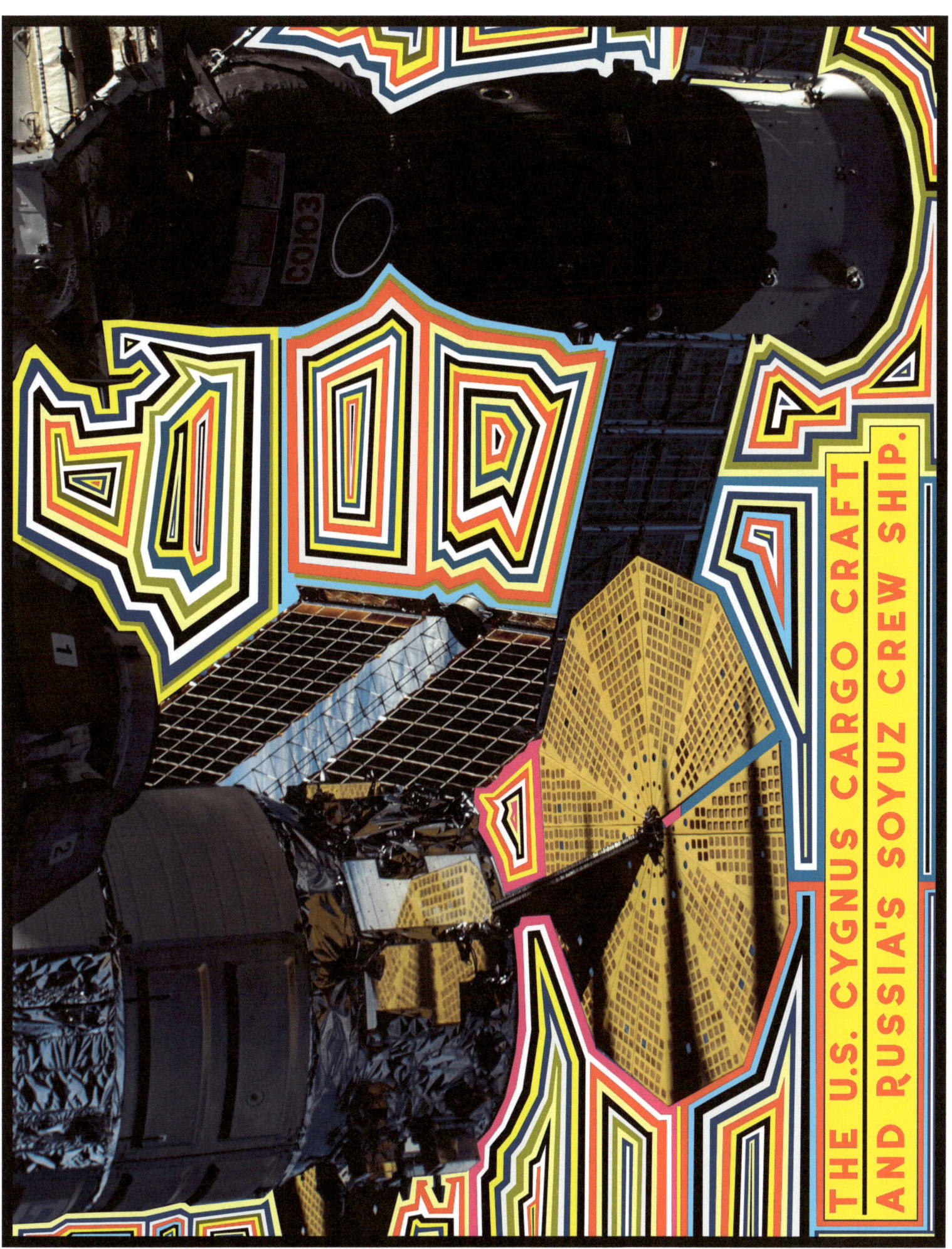

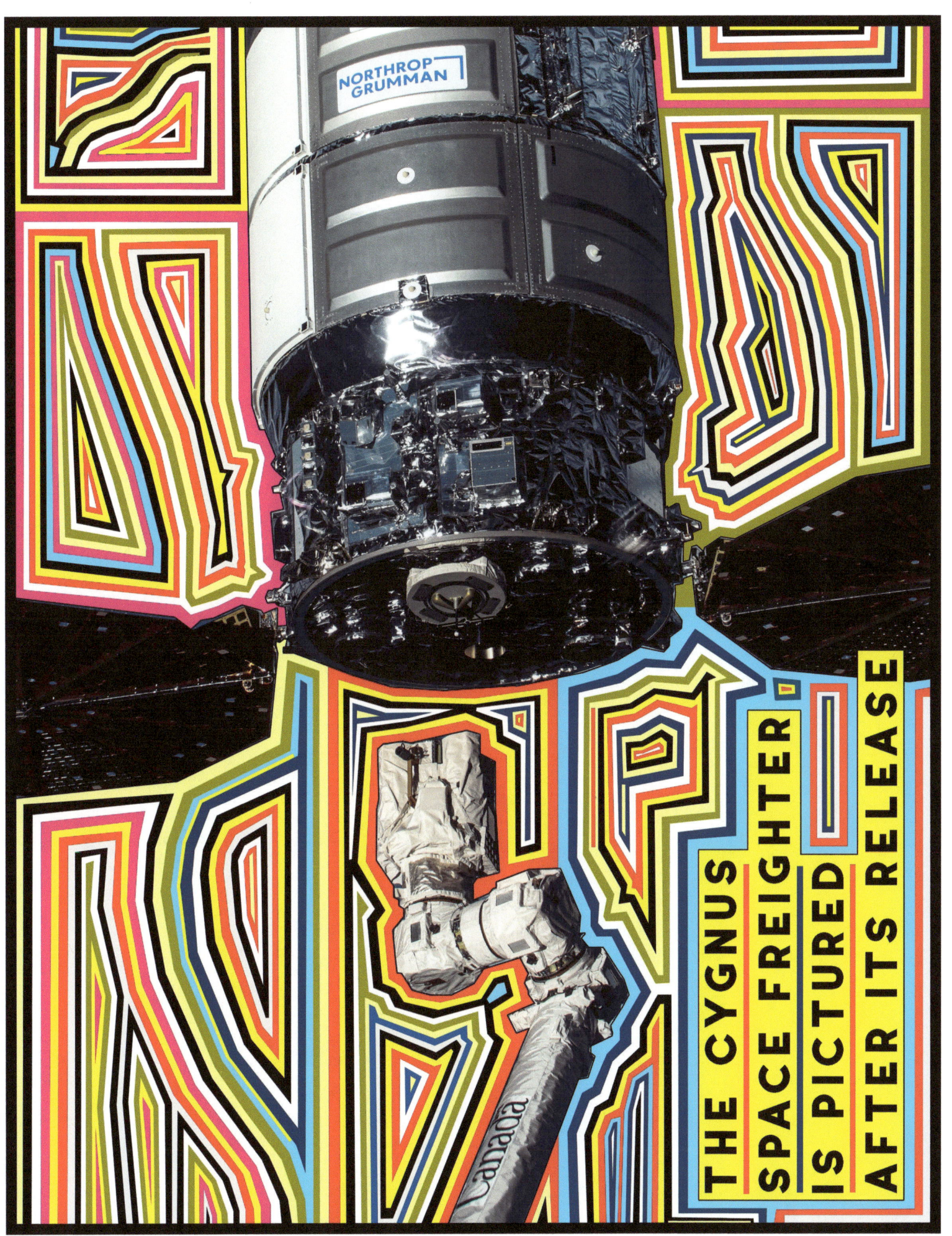

THE CYGNUS SPACE FREIGHTER IS PICTURED AFTER ITS RELEASE

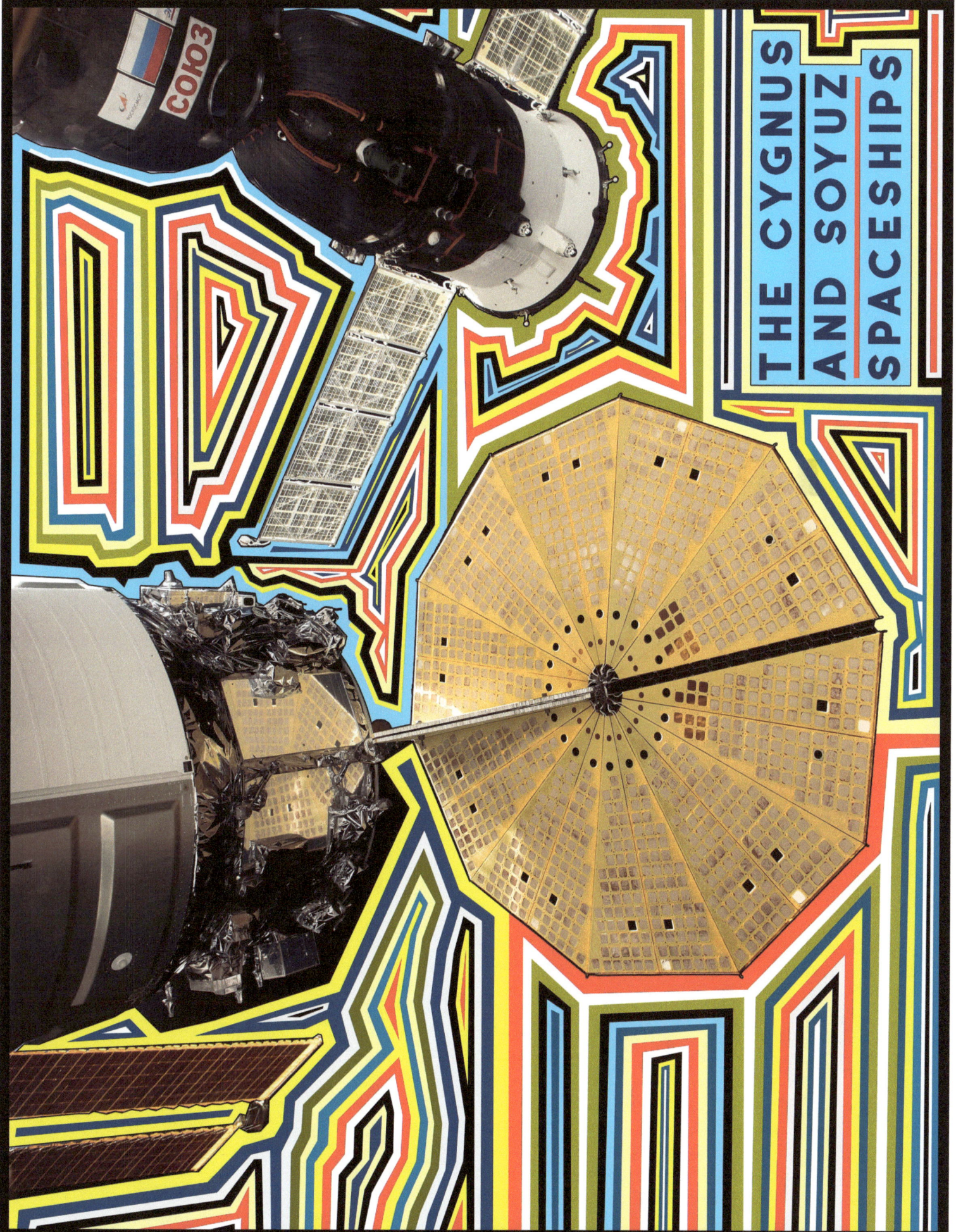

THE CYGNUS AND SOYUZ SPACESHIPS

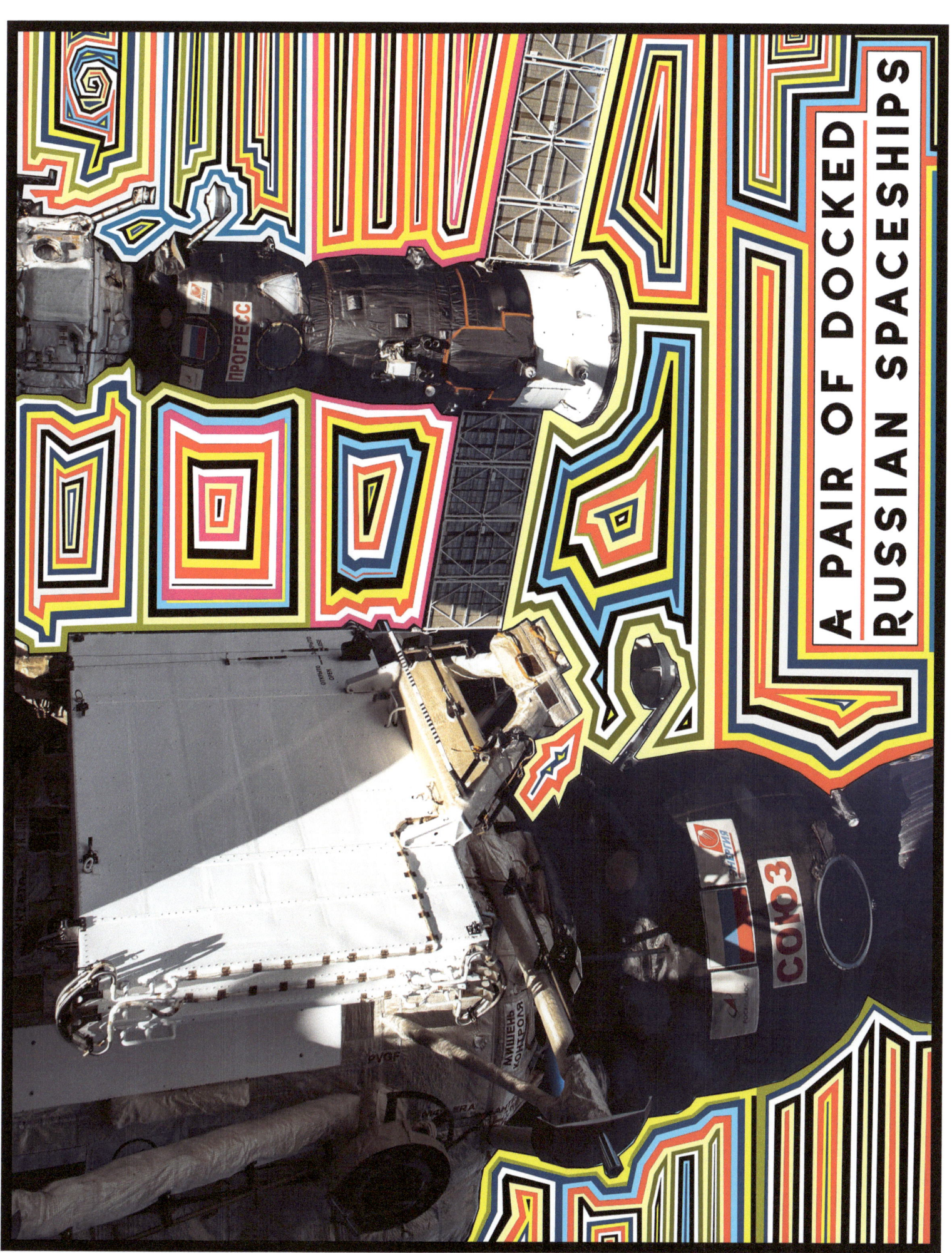

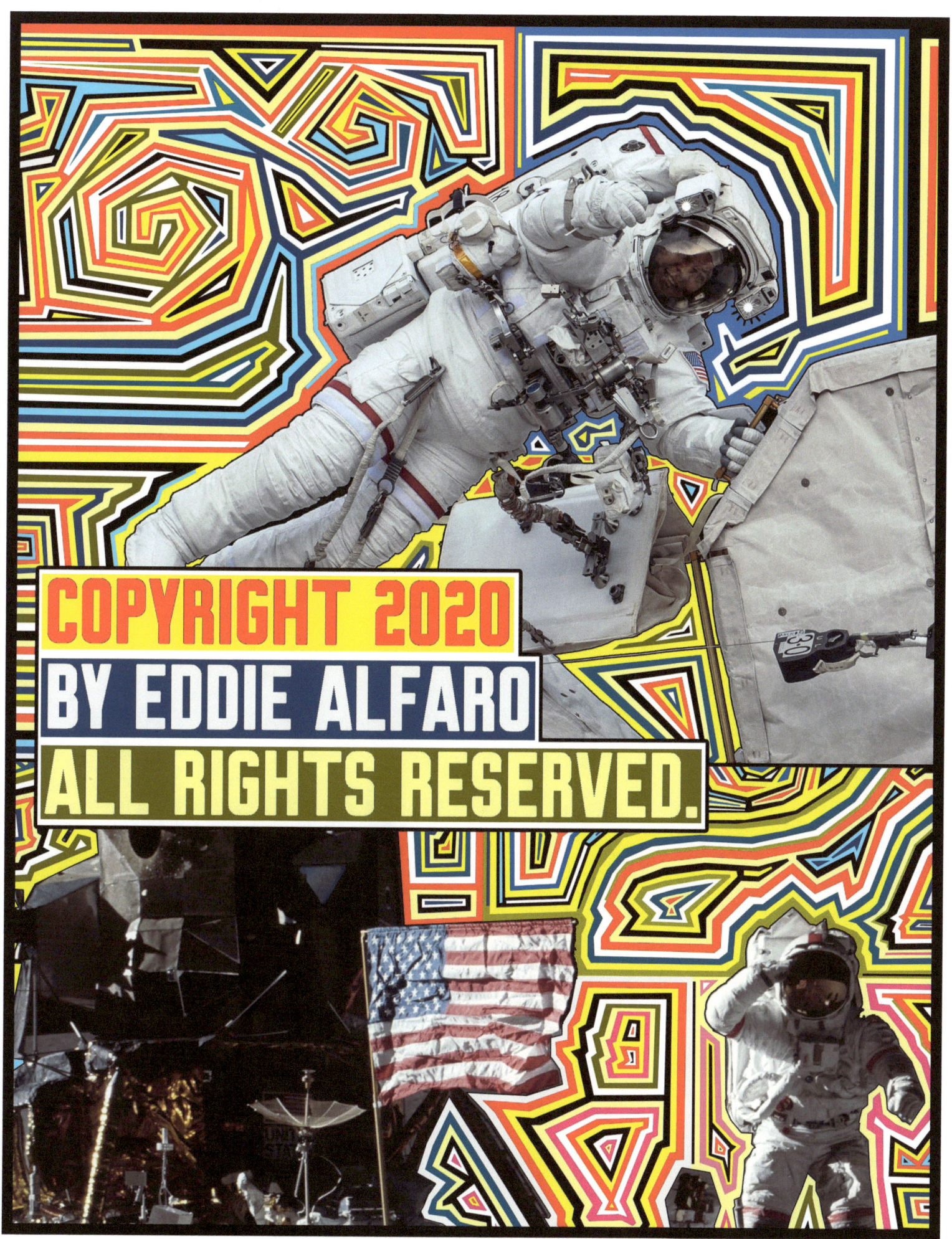

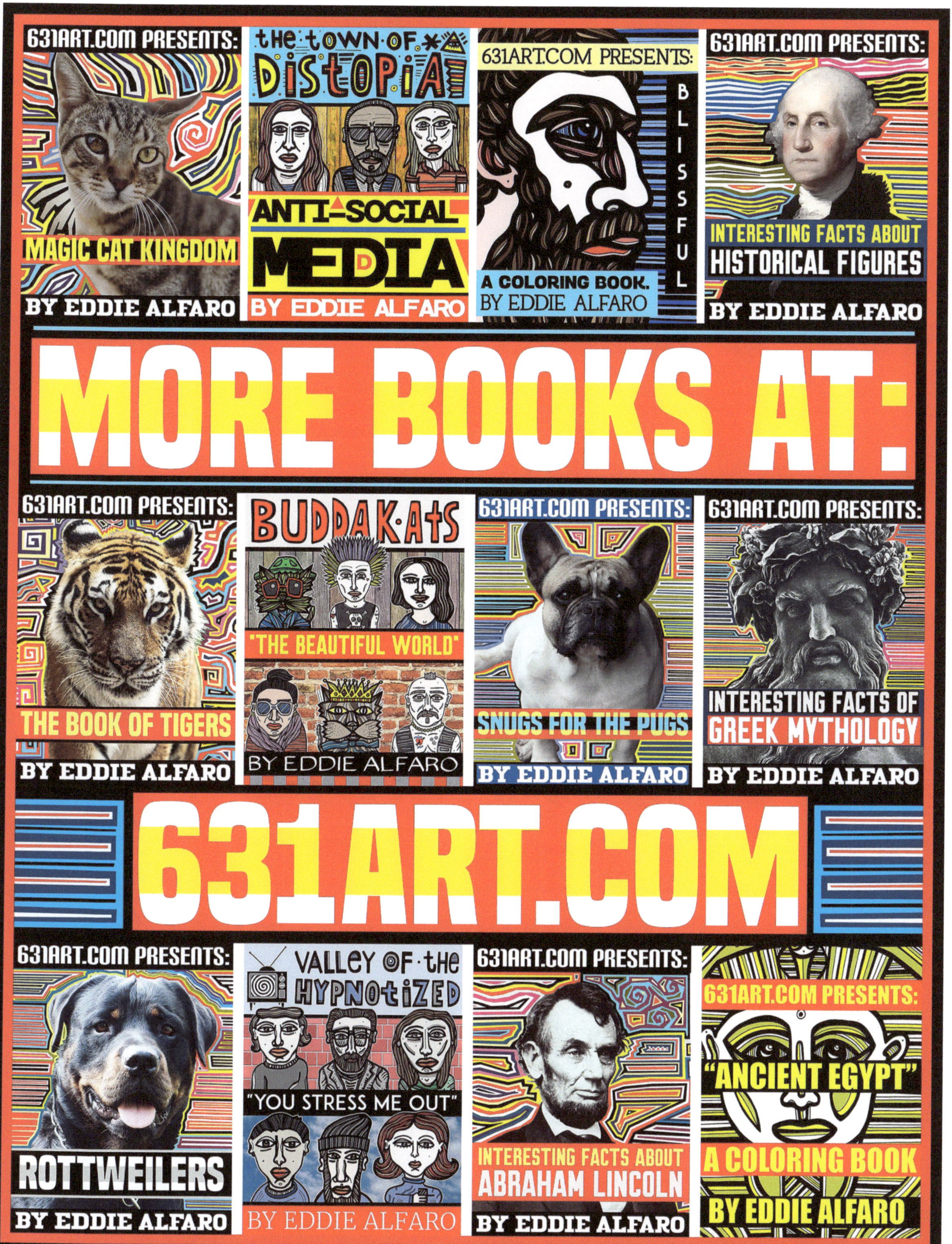
631ART.COM PRESENTS:
MAGIC CAT KINGDOM
BY EDDIE ALFARO
the town of DISTOPIA
ANTI-SOCIAL MEDIA
BY EDDIE ALFARO
631ART.COM PRESENTS:
BLISSFUL
A COLORING BOOK.
BY EDDIE ALFARO
631ART.COM PRESENTS:
INTERESTING FACTS ABOUT
HISTORICAL FIGURES
BY EDDIE ALFARO
MORE BOOKS AT:
631ART.COM PRESENTS:
THE BOOK OF TIGERS
BY EDDIE ALFARO
BUDDAKATS
"THE BEAUTIFUL WORLD"
BY EDDIE ALFARO
631ART.COM PRESENTS:
SNUGS FOR THE PUGS
BY EDDIE ALFARO
631ART.COM PRESENTS:
INTERESTING FACTS OF
GREEK MYTHOLOGY
BY EDDIE ALFARO
631ART.COM
631ART.COM PRESENTS:
ROTTWEILERS
BY EDDIE ALFARO
VALLEY OF the HYPNOtIZED
"YOU STRESS ME OUT"
BY EDDIE ALFARO
631ART.COM PRESENTS:
INTERESTING FACTS ABOUT
ABRAHAM LINCOLN
BY EDDIE ALFARO
631ART.COM PRESENTS:
"ANCIENT EGYPT"
A COLORING BOOK
BY EDDIE ALFARO

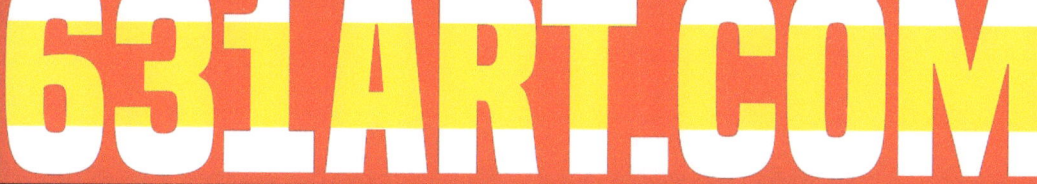

www.ingramcontent.com/pod-product-compliance
Lightning Source LLC
Chambersburg PA
CBHW040410220526
45473CB00004B/1189